ICME-13 Topical Surveys

Series editor

Gabriele Kaiser, Faculty of Education, University of Hamburg, Hamburg, Germany

More information about this series at http://www.springer.com/series/14352

Ghislaine Gueudet · Marianna Bosch
Andrea A. diSessa · Oh Nam Kwon
Lieven Verschaffel

Transitions in Mathematics Education

 Springer Open

Ghislaine Gueudet
CREAD
ESPE Bretagne UBO
Rennes
France

Oh Nam Kwon
Seoul National University
Seoul
Korea, Republic of (South Korea)

Marianna Bosch
IQS School of Management
Universitat Ramon Llull
Barcelona
Spain

Lieven Verschaffel
Van den Heuvelinstituut
Katholieke Universiteit Leuven
Leuven
Belgium

Andrea A. diSessa
University of California
Graduate School of Education
Berkeley, CA
USA

ISSN 2366-5947   ISSN 2366-5955   (electronic)
ICME-13 Topical Surveys
ISBN 978-3-319-31621-5   ISBN 978-3-319-31622-2   (eBook)
DOI 10.1007/978-3-319-31622-2

Library of Congress Control Number: 2016935974

Printed on acid-free paper

This Springer imprint is published by Springer Nature
The registered company is Springer International Publishing AG Switzerland

Contents

Main Topics You Can Find in This ICME-13 Topical Survey

- A review of the kinds of transitions investigated by research in mathematics education and of the associated perspectives
- Conceptual change and learning processes as transition processes
- Transition from university mathematics to teaching mathematics at secondary school and double discontinuity
- Institutional transitions
- Transitions between out-of-school and school mathematics.

Introduction

"Transition" from one state to another (e.g., from childhood to adulthood) is a process of change. Within an educational context change happens everywhere (or at least it should do so); thus transitions are similarly everywhere. Learning can be described as a process of transition (i.e., "the process of transition from a novice's state to that of an expert," Nesher and Peled 1986). Moreover, any change or transition process can be (or seem) either continuous (or discrete but composed of successive steps to be accomplished) or discontinuous, with identified ruptures or gaps.

In this survey we mainly address two kinds of changes: (a) conceptual change and learning processes as transition processes and (b) transitions between social groups or contexts with different mathematical practices.

We propose a state of the art of research review on both themes. This review concerns learning mathematics at all ages: from preschool to university and in the workplace (which can be a school when students become themselves teachers!). The research in this literature review develops various theoretical approaches and associated methods. Most of these raise the issue of continuity or discontinuity in the transitions, sometimes with different conclusions. Work focusing on discontinuities often identifies difficulties attached to these discontinuities, and sometimes it suggests ways to overcome these difficulties.

This survey begins with a literature review ("Survey on the State of the Art"), describing the main perspectives on transition present in the research in mathematics education: the epistemological, cognitive, and sociocultural perspectives. Each perspective has its own characteristic questions and results. Then we deepen these different perspectives in four sections. Section 2 concerns learning processes in the case of difficult concepts. We introduce a view of conceptual change as involving specific "pieces and processes," using the framework of Knowledge in Pieces. Section 3 focuses on the double discontinuity between secondary school mathematics and university mathematics, as introduced by Klein (1908/1939). Students entering university to learn mathematics experience a first transition; if they become secondary school teachers at the end of their studies, they meet a second transition. Section 4 introduces an institutional perspective, viewing mathematical practices as shaped by the institution where they take place. Section 5 considers two kinds of transitions and corresponding research: the transition between prior to school and school mathematics, and the transitions (in both directions) for students between school and out of school mathematics. Finally we describe some directions for future research on transitions in mathematics education.

Survey on the State of the Art

1 What Transitions in Mathematics Education? Review of the Literature

In this section our aim is to set the scene for the following four sections and situate them within the broad picture of mathematics education research concerning transitions. What kinds of transitions have been considered by mathematics education research? What research questions were studied and which theoretical approaches and associated methods were used? Did the studies lead to the identification of continuous processes, successive steps, or discontinuities? Are difficulties attached to the discontinuities identified, and does the research propose means to reduce those difficulties in order to foster a transition? These are the questions we study in this short literature review.

1.1 Mathematics Education: A Story of Change

We do not intend here to build a complete list of the kinds of changes considered by researchers, but to identify central directions for our review. Yerushalmy (2005) introduces a notion of *critical transitions*:

> Critical transition is viewed as a learning situation that is found to involve a noticeable change of point of view. This change could become apparent as an epistemological obstacle, as a cognitive discontinuity or as a didactical gap. A transition would be identified as a necessity for entering into a different type of discourse (in terms of the language, symbols, tools and representations involved) or more broadly as changing "lenses" used to view the concept at hand (p. 37).

© The Author(s) 2016
G. Gueudet et al., *Transitions in Mathematics Education*,
ICME-13 Topical Surveys, DOI 10.1007/978-3-319-31622-2_1

In this definition several possible kinds of transitions appear, all of them concerning a learning situation and a change of point of view: epistemological, cognitive, or discursive. In the literature referring to transitions, we observe all these possibilities and even more.

The changes studied can happen within the mathematical content. Some researchers have studied these changes from an epistemological perspective, sometimes drawing on the history of mathematics. Research from this perspective for the most part also studies transitions in students' learning since transitions in content that are not reflected in student learning are less important to instruction. These cognitive transitions can be studied with different theoretical frameworks; they can concern specific mathematical topics (for example, the transition from arithmetic to algebra has been extensively researched) but also more general issues, such as the transition between different thinking modes. We present and discuss the corresponding research works in Sect. 1.2.

Other authors consider that mathematics is shaped by groups of people who develop shared mathematical practices. Educational institutions are such groups: much research considers the transition, for example, between primary and secondary school. But mathematical practices also exist outside of school, and the transition between mathematical practices in school and in the workplace have also been investigated, mainly by researchers with a sociocultural perspective. These studies often identify discontinuities and sometimes design teaching experiments in order to smooth the transition. Research works that have a sociocultural perspective are presented in Sect. 1.3. Some authors argue that "transitions can be best understood using a socio-cultural framework" (Crafter and Maunder 2012). Our aim here is not to discuss whether one or the other perspective is more appropriate, but to show that these perspectives act as different lenses on the issue of transition (Gueudet 2008), and to depict the main features of the results they produce (naturally, sometimes the same transition has been studied with multiple perspectives, such as the arithmetic-algebra transition (e.g., Goodson-Espy 1998; Sadovsky and Sessa 2005; Slavit 1999).

Other kinds of transitions have been studied within mathematics education research: for example, the changes in the available tools to do mathematics are also associated with transitions. For example the transition from using tracers and templates to using a compass for drawing circles, for students (Chassapis 1998) or the transition experienced by teachers moving towards CAS-supported classrooms (Kendal and Stacey 2002) have been studied by researchers. We did not include the change from a given tool to another as a central kind of transition for this literature review. Our interest in tools and technologies is more directed towards how the available tools impact a given transition: the availability, for example, of dynamic geometry software can impact students' learning of deductive geometry, hence modifying their transition processes between intuitive and deductive geometry. This is Yerushalmy's (2005) perspective, who gives evidence that some known transitions have a different nature in technological environments.

1.2   Epistemological and Cognitive Transitions

The history of mathematics is also a story of changes, of long-term genesis of concepts with moments of continuity and sudden ruptures. How history can inform research in mathematics education is a complex matter (Schubring 2011); here we consider how research in mathematics education has used the history of mathematics in studying transitions issues. For example, Dorier (2000) studies the genesis of linear algebra and shows that this theory has emerged in particular from a need of unification of different problems concerning functions, sequences, and so forth. Thus linear algebra is what Robert (1998) calls a formalizing, unifying, and generalizing theory: it cannot be taught as a "natural" extension of previous contents or a solution to a given problem. Studying the history of mathematics draws a landscape with gaps and long paths that do not admit shortcuts. An "epistemological obstacle" is a form of unavoidable discontinuity which has been studied by many researchers (e.g., Sierpińska 1987), most often by identifying first such obstacles with a historical perspective then studying the associated cognitive transitions.

Cognitive transitions are also observed outside of specific mathematical contents. Concerning the learning of proof in particular, several kinds of transitions have been investigated. In the transition from phases of conjectures to phases of proof continuities and discontinuities appear, and this leads to the introduction of the concept of cognitive unity of theorems (continuity existing between the production of a conjecture and the possible construction of its proof; see Garuti et al. 1996). Other researchers have identified structural discontinuities in the transition between argumentation and proof (Arzarello and Sabena 2011).

General models of transitions taking place in mathematical thinking also exist; in particular, different theories propose models for the "transition from process to object" (Tall et al. 1999). This transition is sometimes called "encapsulation" and sometimes "reification"; several authors claim that it is composed of a sequence of steps. Sfard (1991) considers the transition from computational operations to abstract objects to be accomplished in three steps: *interiorization, condensation*, and *reification*. Dubinsky (1991) in the APOS theory also considers conceptualization processes as transitions composed of three successive steps: from action to process, from process to object (encapsulation), and from object to schema. Each of the steps, or sub-processes, is seen as continuous, while the change from one step to the next can be interpreted as a discontinuity. Nevertheless, the process-object transition is not discussed by these authors in terms of continuity/discontinuity. In contrast, Tall (2002), referring to Skemp (1962) for a study of "long-term learning schemas" discusses the process-object transition in terms of discontinuity. The author claims that long-term cognitive development in mathematics always faces discontinuities. He gives various examples concerning negative numbers, algebra, limits, and so forth. These discontinuities, according to him, cannot be avoided in the curriculum, and they require cognitive reconstruction on the part of the student. It is thus important for teachers to be aware of these discontinuities. Should we then maintain that cognitive development is generally a discontinuous process or merely a concatenation of continuous sub-processes? This issue is discussed in Sect. 2.

1.3 Sociocultural Transitions

Researchers adopting a sociocultural perspective have also studied transitions. Different groups of people share a common mathematical practice; a transition happens when an individual, such as a student, moves from one group to another. The research questions can address the characteristic features of the relevant mathematics and mathematical practices in each group: What are these character-istic features? What are the similarities (continuity) and differences (discontinuity) between the mathematics and mathematical practices of different groups?

The more precise formulation of the questions, the methods employed, and the results obtained depend on several factors, including the groups considered. Those groups can be of different natures. For example, the ICMI Pipeline project (Hodgson 2015) whose aim was to "study issues associated with the supply and demand for mathematics students and personnel in educational institutions and the workplace" focused on what they called four "transition points":

1. School to undergraduate courses,
2. Undergraduate courses to postgraduate courses,
3. University into employment, and
4. University into teaching.

In this case we observe two kinds of transition points: between two schooling institutions (which are here two successive levels) and between a schooling insti-tution (namely university) and a workplace, with a distinction between a general workplace (employment) and working as a teacher. Section 3 in this survey con-cerns this last case.

Other groups are considered in other research: transition between mathematics teaching in different languages (Ríordáin and O'Donoghue 2011) and between mathematics learned at school and at home (Crafter and De Abreu 2011). We notice that these two last kinds of groups can lead to a different sort of transition. While transition from primary to secondary school is a non-reversible process corre-sponding to a temporal progression, students can simultaneously take courses in two languages and can certainly experience their families and schools simultane-ously. These continuous interactions between different contexts can be seen as *permanent transitions* between two aspects of a student's mathematical practices; here continuity means a kind of connectedness.

Which are the theoretical approaches used to study the transitions between all these different groups? We here briefly survey the main approaches and associated work.

Some research employs the anthropological theory of didactics (ATD, Chevallard 2006), which posits that institutions shape mathematical practices. ATD has mostly been used to research transition between different school levels (e.g., Winsløw 2014). Section 4 presents a survey of work using this approach.

Other researchers focus on the change in the mathematical discourse between different groups. The commognitive approach (Sfard 2007) is such a discursive approach, which considers mathematical learning as initiation into—or transition to—a mathematical discourse, characterized by *word use*, *visual mediators*, *endorsed narratives*, and *routines* shared by a community. As with ATD research, these characteristics can be identified for a given community and compared with another, and the acculturation of a student can be followed. Such research can also identify students' difficulties. However, in contrast to ATD, this approach probably offers a more continuous perspective on transition that is less focused on discontinuities or gaps.

The concept of community has been especially developed within the theory of communities of practice (CoPs, Wenger 1998), which has also been used to investigate transitions. According to this theory, learning takes place through interactions within a community sharing a common practice; it also takes place in the evolution from a legitimate peripheral participation to a full participation as member of the community. For example, first-year students at university participate in the mathematics undergraduate student community, which has its own rules for studying and communicating about mathematics (Biza et al. 2014). The process of transition corresponds here with the process of becoming a member of this community, and the theory of CoPs interprets it as a learning process. In the theory of CoPs, it is also common for a subject to belong simultaneously to two different communities. In this case of co-existing communities, the concepts of boundary crossings and boundary objects permit understanding the permanent transitions experienced by the subject (see e.g., Crafter and de Abreu 2011): they constitute the continuous aspect of these transitions. Such perspectives are used to study in particular the transitions between school and out-of-school mathematics, which are the focus of Sect. 5.

The following sections propose a more detailed view of the questions studied and of the results obtained by selected mathematics education research on transitions using diverse perspectives.

2 Continuity Versus Discontinuity in Learning Difficult Concepts

2.1 Setting the Scene

This brief survey attends to a certain kind of transition—that between naïve or novice knowledge and the knowledge of an expert. My perspective is largely cognitive, hence learning is construed as change in conceptual structure. In addition, I focus on one particular issue concerning transitions in conceptual structure, whether they should be regarded as continuous or discontinuous (abbreviated as CvD—Continuous vs. Discontinuous).

CvD has had an interesting history in mathematics education. A primary locus of attention is the idea of "epistemological obstacles" in learning. Epistemological obstacles are unavoidable difficulties in learning that manifest as persistent wrong or inappropriate interpretations by learners. Simple examples include the non-sensibility of negative numbers (what can negative four cows mean!), and, in more advanced mathematics, a slew of misconceptions about the concept of limit (e.g., confusion between an infinite set and properties of a never-ending process: Zeno's Paradox—a hare can never catch a tortoise because it has to endlessly half the distance to the tortoise). Researchers such as Tall (e.g., 2002) and Sierpińska (e.g., 1990) have advanced the idea of epistemological obstacles in mathematics learning. Broadly, I believe it is fair to say that the idea of epistemological obstacles favors discontinuity in learning; "ruptures" or "breaks" occur precisely at the point of surpassing obstacles.

Before returning to mathematics, I will take a detour through the history of CvD in science education. I do this because (1) I know science better, and (2) some important issues concerning CvD are better highlighted in science education, and (3) some CvD-focused empirical strategies are better represented in science.

2.2   Misconceptions and Discontinuities

In the early 1980s, science education researchers noted the striking phenomenon of "naïve conceptions." Students are not blank slates when it comes to learning important topics such as Newton's laws (mechanics), the focus of much early study. Instead, students possess intuitive ideas that were seen almost exclusively as blocks to learning. In fact, the most visible point of view was that there is an essential gap or discontinuity between naïve ideas and normative ones.

A philosophical debate about the nature of science preceded the debate among educators. That debate defined a lot of the terrain for later discussions, including those in education. The far more famous position was taken by Kuhn (1970) in his book on "scientific revolutions." Everyone agrees that creating new science—or, in the educational case, achieving deep understanding of existing science—is difficult and may take a long time. But how should one view the transition and the reasons for difficulty? Kuhn's position was that science before a revolution is flatly discontinuous ("incommensurable") with that afterward. The reason for difficulty of change is not just in the differences between views before and after the revolution, but that science, at any point in time, is a coherent whole. If change is to happen, basically everything must change at once. He likened the change to a gestalt switch, such as illustrated in the "duck/rabbit," which Wittgenstein used in his philosophical investigations (Fig. 1). The "duck/rabbit" can be seen as either a duck or as a rabbit, but not both at the same time. Changing from one interpretation to the other constitutes a holistic change.

Kuhn's less well-known competitor was Toulmin (e.g., 1972). Toulmin made two central points. First, he believed that science is much less coherent than Kuhn

Fig. 1 The "duck/rabbit"
used by Wittgenstein

assumed; he disparaged Kuhn's ideas as those of a "cult of systematicity." Gestalt switches are thus not plausible. Second, he observed that before/after "snapshot" studies of science are very likely to promote an illusion of discontinuity. To make an analogy, one is very likely to see a pile of materials—bricks, boards, nails, bags of cement—as categorically distinct from the house constructed of those materials. The way to understand continuity is to imagine the intervening small steps of construction, which Toulmin described as a "moving picture" account.

A comparable philosophical pre-history in mathematics involves Bachelard (1938). He developed the ideas of "epistemological obstacle" and "epistemological rupture" that were picked up by Tall, Sierpińska and others. Epistemological obstacles in education are roughly the equivalent of "misconceptions" in science. While Kuhn and Bachelard's ideas seem very similar with respect to CvD, historical accounts suggest they were developed independently.

2.3 Conceptual Change

Carey (e.g., 1985) brought Kuhn's ideas forcibly into the study of difficult learning, called "conceptual change." To this day, Carey's work (e.g., Carey 2009) remains a landmark in this field. And other prominent current work shares assumptions with Carey, although in perhaps weakened form. Vosniadou (e.g., 2013) moves the real resistance to learning from the misconceptions themselves to "framework theories," which lie in the background but strongly constrain conceptions; framework theories embody our deepest assumptions about how the world works. Despite differences between her and Carey, Vosniadou is still committed to fairly strong assumptions of coherence in students' ideas (for a review, see diSessa 2013). Despite its importance in the CvD debate, the issue of coherence is much less visible in some other modern views of conceptual change (e.g., Chi 1992, 2013).

Even while Carey was championing Kuhn, other researchers took Toulmin's side, favoring continuity. Minstrell (1989), for example, likened learning to selecting and weaving naïve strands of thought into a different, more systematic fabric. A line of conceptual change study known as Knowledge in Pieces (KiP; diSessa 1982) began by identifying, both theoretically and empirically, some of Minstrell's strands of understanding that need to be woven or rewoven. At that time, despite this competing research, claims of high systematicity and gestalt switches (ruptures and revolutions) were dominant.

2.4   Steps Toward Resolving the Conflict

To resolve the contest between gestalt switches and moving picture accounts, one needs two things: (1) One needs to understand the pieces involved in the change. For the building-a-house case, one needs to know about nails, boards, and so on. (2) Then one needs an array of processes that can transform those materials into a house. Nails bind boards; cement, with added water, transforms itself gradually into a hard and shape-permanent material. Unfortunately, and perhaps shockingly, careful accounts of the "materials" involved in conceptual change—much less empirically precise tracking of processes involving them—has been and is still rare. Much more common are before-and-after studies, with little or no process data in between. The relatively recent trend toward microgenetic (moment-by-moment) studies of learning (Parnafes and diSessa 2013) shows great promise in answering Toulmin's call for "moving picture" accounts of conceptual change, and therefore definitively settling the CvD dispute: Will we see revolutions and ruptures, or not?

Since it highlights a search for pieces and processes and attempts to build detailed "pieces and processes" accounts of learning, contemporary KiP research is positioned well to help resolve the CvD dispute. I mentioned before that early work in the KiP line developed accounts of relevant pieces (naïve knowledge). More recently, theoretical and empirical emphasis has shifted to the processes of conceptual change. The conceptual equivalent of pounding nails, mixing cement, and intermediate and temporary constructions such as scaffolding may be coming into view.

Prominent advocates of continuity and "pieces and processes" research are rarer in mathematics education research compared to science, which is one of the main reasons for the diversion to science. Unfortunately, space does not permit a comparative review.

2.5   Empirical Work

Here are two brief descriptions of recent "pieces and processes" (microgenetic) studies of learning and conceptual change.

2.5.1   The Construction of Causal Schemes

This work (diSessa 2014) is based on an early KiP study that identified intuitive knowledge elements behind "misconceptions" in science learning. In particular a subset of those previously identified elements (two key ones, and a few that are more peripheral) became involved in the learning event at issue.

That learning event seemed, in some respects, remarkable. A class of high school students—with minimal scaffolding, no direct instruction, and in less than an hour's time—managed to construct on their own a version of the targeted scientific idea, "Newton's laws of thermal equilibration": Two objects in contact each change temperature at a rate proportional to the difference in temperatures between them. Six independent mechanisms (basic processes) were identified in the case study. Several of these processes seemed "discrete," such as developing a chain of causal links (A causes B; B causes C; …), the totality of which was precisely the causality behind Newton's thermal laws. The chain was primed by one previously identified intuitive idea, and contained at its core another very important and previously identified intuitive causal link. Other parts of the construction seemed less "logical and scientific." For example, the central intuitive causal link was expressed in overtly anthropomorphic ways ("the objects in contact want to be in equilibrium, and are freaking out because they are so far apart"). Remarkably, the anthropomorphism ("wanting" and "freaking out") gradually disappeared from the students' talk, and, in the end, they had a very professional-sounding version of Newton's laws: The rate of temperature change is proportional to the difference in temperatures. The anthropomorphism seemed to be a scaffold and not a permanent part of the construction, just like scaffolds are critical, but they come and go in the construction of a house (and may, thus, never be seen in before-and-after studies!).

2.5.2 A Mathematical Example: The Law of Large Numbers

I mentioned earlier that "pieces and processes" study of mathematical learning is rarer than for science. This example is an exception. (For a related example, see Pratt and Noss 2002.) While it is too early to generalize to all of mathematics learning the example suggests that, at the microgenetic grain-size, there is no substantial discontinuity in learning.

Wagner (2006) used clinical interviews to study learning of the statistical law of large numbers. One of the theoretical tools developed by KiP shows how concepts may not be monolithic, but actually composed of many pieces each of which works only in certain contexts. Then, learning may proceed incrementally, one context at a time. In Wagner's study a student gradually linked in intuitive schemes (similar to the science example, above) that were approximations of normative ideas. The idea that "larger samples are more accurate" occurred to her specifically in the context of surveys, and demonstrably changed the range of contexts in which she could effectively apply the law of large numbers. In general, context-specific intuitive ideas seemed necessary to extend her capacity to apply the law of large numbers across new contexts, and it appeared that linking those ideas in was distributed in time, with no categorical breakthroughs (gestalt switches).

2.6 Implications for Teachers and Other Educational Professionals

Let me discuss a list of properties that describe what I take to be key components of the misconceptions/revolutions/epistemological obstacles points of view, in contrast to a more continuist point of view.

Misconceptions/epistemological obstacles are:

- *Manifest* as persistent wrong or inappropriate interpretations by students.

But, recognizing a large number of intuitive pre-existing ingredients, many of them will find "happy homes" in more advanced scientific thinking. Both the scientific and mathematical case studies, above, found this to be true in specific details of learning. Although it tends not to be noticed, consistently easy accomplishments, not just blocks, arise because of the existence of pre-instruction knowledge. We need most definitely to study the general nature and specifics of "things that may help," not just mark occasions where help may seem beyond our grasp.

- *Unavoidable.*

There are many routes to constructing a house. What was found with thermal equilibrium is a very different avenue of learning than that stipulated by Carey and colleagues (e.g., Wiser and Carey 1983; Wiser 1995), based on assuming a coherent naïve conception of thermal phenomena and consequent revolutionary changes in conceptions. In general, unless we have an encompassing map of potentially relevant naïve knowledge, it is unwise to assume that we cannot find more productive pathways from naïve to expert understanding than we know at present.

- *Monolithic.*

All conceptual change researchers agree that some things are difficult to learn and thereby take time. But, consider: If you don't know how to do it, building a house may seem almost impossible. Even knowing the ingredients (boards, nails, cement)—which may not at all be securely in place for conceptual change—may not be enough. When one gets down to pounding nails, one by one, however, one can see how many little steps can accumulate. One promise of continuist approaches to learning is that we may always need time to teach difficult ideas, but the path may not need to be opaque or painful.

- *Homogeneous.*

Studies such as those above suggest that the ingredients and processes of change may be different from concept to concept. Cases of "overcoming obstacles," therefore, may also look very different from one another. Calling everything an "epistemological rupture" (or a "naïve theory" or even a "misconception") paints change with too broad a brush so that it is unhelpful in guiding us toward success in individual cases of conceptual change.

To sum up, the continuist point of view sees learning and conceptual change as complex, but open to possibilities that have simply not been pursued very much, owing to the historical dominance of discontinuity as a core assumption. From the continuist perspective, we can (we must!) explore more widely to search out productive paths to expertise. Even if some blocks to learning are fairly characterized as "obstacles," they may be better conceptualized as long journeys, rather than walls to scale right here and right now, or chasms to somehow jump over.

2.7 Final Words

While it is tempting to believe that the CvD dispute might be settled by which community can produce the best instruction, I believe that the issue is best construed as, and will only be settled as, a scientific difference of opinion. Instruction is too complex an affair (respecting cognitive, but also social, affective, cultural, and institutional issues) to serve as a likely arbiter in the short term. As for making progress on the science, as I have anticipated, my view is that what is necessary is better understanding of the diverse elements that learners bring to bear in their learning, and also in much more and more careful micro-analyses of the intermediate steps and learning mechanisms.

3 Double Discontinuity Between Secondary School Mathematics and University Mathematics: Focusing on Mathematical Knowledge for Teaching

The transition from secondary school to university and the backwards transition from university to secondary school are challenges for school teachers in their life trajectories to becoming professional teachers. Each transition involves changes of the teaching-learning culture and in the relevant type of mathematics. The challenges of these bi-directional transitions are what Felix Klein (1908/1939) referred to as a "double discontinuity."

The first discontinuity deals with the well-known problems that students face as they enter university, a main theme in research on university mathematics education (e.g., Gueudet 2008). The second discontinuity deals with those who return to school as a teacher and the transformation of disciplinary mathematics at university to school mathematics. In general, these discontinuities can be traced back to the difference between the mathematical paradigms prevalent at school and university. Even nowadays, the two discontinuities still seem to persist, and future teachers often believe that the topics of university mathematics simply do not fit the demands of their later profession in school. Winsløw and Grønbæk (2014) distinguished three dimensions of Klein's double discontinuity: the institutional context (of university vs. school), the difference in the subject's role within the institution (a student in university or school vs. a teacher of school mathematics), and the difference in mathematical content (scientific vs. elementary). Bosch (this volume, Sect. 4) discusses aspects of the institutional context for the forward transition from secondary to university. This section aims to provide an overview of the current state of the art in the context of teacher education from an international perspective in order to provide a deeper understanding of double discontinuity phenomena, with a special focus on mathematical knowledge. Specifically, we maintain that tax-onomies of mathematical knowledge for teaching constitute attempts to locate, characterize, and say exactly why there is a discontinuity between university learning and teachers' experience returning to schools.

3.1 Theorizing Teacher Knowledge

Understanding teacher knowledge of mathematics and its impact on teaching and learning is difficult and has become increasingly difficult as the complexity and dynamic nature of knowledge has become recognized. The theoretical constructs of teacher knowledge provide lenses through which one can examine research about learning to teach in the context of the bi-directional transition between university and school. At a fundamental level, this understanding concerns the production of a set of criteria for what constitutes the professional knowledge of mathematics teachers and what constitutes the essential features of mathematics teacher educa-tion curricula. In this respect, Shulman's (1986) seminal work on teacher knowl-edge provides a useful guide on how to expand previous content knowledge-oriented teacher education curricula. He proposed a framework for analyzing teachers' knowledge that distinguished different categories of knowledge. Among these categories, Shulman emphasized pedagogical content knowledge, "subject matter knowledge for teaching" (Shulman 1986, p. 9), as being important because it identifies the distinctive ways of understanding mathematics for the purpose of teaching. The distinction between general pedagogical knowledge (PK), content knowledge (CK), and pedagogical content knowledge (PCK) has proved practically useful and has been employed in numerous studies (e.g., Borko 2004; Blömeke et al. 2011). However, the assessment of teachers' CK and PCK requires a theory of the subject in question and of its knowledge. There is broad consensus that these two components of professional knowledge cannot simply be equated with a command of the material taught. Nevertheless, theory-driven approaches to the assessment of teachers' CK and PCK remain rare.

Some conceptions have, however, been developed and empirically tested for the subject of mathematics. Ball et al. (2005, 2009) and Hill et al. (2008) have developed a theoretical framework and empirical measures to assess professional content knowledge that mathematics teachers need in order to teach effectively. Their frame of reference is not university-level knowledge but the mathematics behind the institutionalized curriculum of elementary school mathematics. On this basis, Ball et al. (2005) and Hill et al. (2004) distinguish the everyday mathematical knowledge that every educated adult should have (common knowledge of content) from the specialist knowledge acquired through professional training and classroom experience (specialized knowledge of content). They further identify a third dimension of mathematical knowledge, which links mathematical content with student cognition (including misconceptions and pedagogical strategies), namely, knowledge of students' apprehension and learning of mathematics. At the same time, they distinguish three content areas of elementary mathematics: numbers and operations, patterns and functions, and algebra. Ball's group has used a matrix of these content areas and knowledge dimensions as a theoretical structure for the development of test items, with items being allocated to the cells of the matrix on the basis of prior theoretical considerations. On the basis of these analyses, they developed an overall test based on item response theory assessing elementary school teachers' mathematical knowledge for teaching (MKT). In doing so, they refined the notion of PCK to introduce MKT, which they defined as a specialized knowledge of mathematics situated in the context of teaching. MKT consists of four domains of mathematical knowledge: common content knowledge (CCK), specialized content knowledge (SCK), knowledge of content and students (KCS), and knowledge of content and teaching (KCT). The four domains of MKT can be broadly categorized into subject matter knowledge (i.e., CCK and SCK) and pedagogical content knowledge (i.e., KCS and KCT) (Hill et al. 2008). More importantly, MKT highlights the idea that the knowledge required for teaching is determined by the practice of teaching itself. The Teacher Education and Development Study: Learning to Teach Mathematics (TEDS-M) provided the opportunity to examine the outcomes of teacher education in terms of teacher knowledge and teacher beliefs across countries. The TEDS-M concept of teacher education outcomes is based on the notion of "professional competence" (Blömeke et al. 2014). Competence is defined as those latent dispositions that enable professionals to master their job-related tasks, Teacher knowledge as one facet of competence was conceptualized in TEDS-M (for further details, see Blömeke et al. 2014). There are many studies of teachers' mathematical knowledge for teaching mathematics, mostly at pre-high school levels. A growing number of articles address high school teachers' and university instructors' MKT (e.g., Boston 2012; Herbst and Kosko 2014; Lai and Weber 2013; Lewis and Blunk 2012), but the vast majority of studies are at the elementary level.

A major criticism of research on teachers' MKT is that knowledge, the central construct of MKT, is rarely defined and is therefore operationalized inconsistently across investigations (Thompson 2013). Mason and Spence (1999) introduced these distinctions: knowing-to (act in the moment), knowing-that, knowing-how, and knowing-why (an action is appropriate), arguing that knowing-to and knowing-why are the most important forms of knowing for teachers. Thompson (2015) argued that a focus on teachers' meanings is in line with Mason and Spence's argument and that a focus on teachers' mathematical meanings is profitable for understanding their instructional decisions, both in planning and in moments of teaching. His research group developed the Mathematical Meaning for Teaching Secondary Mathematics, a diagnostic instrument designed to give insights into the mathematical meanings with which teachers operate.

Although the previous studies showed the importance of the assessed knowledge components for teaching quality and student learning, they could not answer important questions concerning the structure of mathematics teachers' knowledge. In particular, the relation between CK and PCK is still unclear: Although these components seem clearly separable from a theoretical point of view, most studies found that CK and PCK are highly correlated and sometimes even hard to separate (Blömeke et al. 2014). However, it is not clear if this strong correlation is caused by the underlying conceptualizations or the chosen operationalization. Although CK is often described as knowledge of disciplinary mathematics acquired through formal teacher education, most operationalization is predominantly focused on mathematical content as it is found in schools. This means in particular that the corresponding tests are not appropriate to measure learning progress in pre-service teacher education. Similarly, PCK is described as a kind of knowledge specific to teaching mathematics, but existing test items are often solvable by analytical mathematical competence that has nothing obviously to do with learning or teaching. In this context, Loch et al. (2015) suggested a three-dimensional structure of pre-service mathematics teachers' domain-specific knowledge. In particular, they conceptualized school-related content knowledge (SRCK) as applying academic mathematical knowledge in the context of school mathematics for teaching purposes, which turned out to be separable from CK and was found to be distinguishable from PCK. SRCK may be interpreted as a link between CK and PCK. Though CK, PCK, MKT, and SRCK are directly addressed in courses in university teacher education programs or in other teacher education institutes, the development of teachers' professional knowledge and the relationship between all these elements of teachers' knowledge and teachers' teaching actions in the class are still not comprehensively understood.

3.2   Developments in Teacher Preparation

There have been calls for rethinking mathematical education for future teachers in many countries (e.g., CBMS 2012; Kwon et al. 2012). Criticism about the inefficiency of teacher education is common in the literature. Buchholtz et al.'s evaluation study (2013) of teachers' knowledge of elementary mathematics from an advanced standpoint indicated that during their university education many future mathematics teachers do not succeed in acquiring the deeper mathematical knowledge needed to dismantle school-related misconceptions and solve elementary mathematical problems competently. Teacher education programs at universities fail to provide student teachers with adequate and in-depth learning opportunities for CK and PCK.

Likewise, teacher education curricula are continually criticized as being insufficient to develop professionalism in pre-service teachers. One criticism is that there is no connection between the teacher education curriculum and actual teaching practice. Some argue that there has been arbitrariness in teacher education programs, often being more like a master craftsman's diploma than a theory-guided education in teaching (e.g., Blömeke et al. 2011). Another criticism has been that the mathematics education curriculum for future secondary mathematics teachers is not essentially different from the curriculum for mathematics majors, as shown in the analysis of a current mathematics teacher education curriculum (Kwon et al. 2012).

In order to overcome these criticisms of teacher preparation, there have been several approaches to bridging the gap between university mathematics and school mathematics. Klein (1933) proposed a course on school mathematics from a higher level that addresses the second discontinuity, i.e., the transition from university to school. Klein's proposal for such a course implies that students already have rather advanced mathematics knowledge and can operate with this knowledge in a flexible and mathematically competent way; in particular, the proposal implies that they do not only know facts and techniques but possess adequate, networked, and critical ideas for validating and evaluating the significance of those ideas and techniques. Nowadays, an average student in an upper secondary teacher program may not show this expertise even after finishing their mathematics coursework. This may be the main reason that courses following Klein's recommendations are seldom proposed today. Instead, courses aimed at supporting student teachers in connecting university and school mathematics (typically run in the first or second semester) focus directly and explicitly on connecting the mathematics experienced in these different environments (Winsløw and Grønbæk 2014). There are two principal approaches to interconnecting these different kind of mathematics, which may be mixed: one that adds aspects of the new university discourse slowly and step-by-step and one that develops university-level problems starting with school mathematics.

Another approach is to develop courses on explicitly integrating CK with PCK in mathematics and the didactics of mathematics (Krauss et al. 2013). These courses provide students the opportunity to reflect on school-relevant mathematical content and thus to construct their individual network of content, pedagogical content, and general pedagogical knowledge. Kaiser and Buchholtz (2014) developed ideas for a program in mathematics teacher education to overcome the gap between university and school mathematics. Their project aimed particularly at first semester student teachers, and it offered mathematics in a way intended to address the first of Klein's discontinuities.

3.3 Final Comments

In sum, Klein's notion of a double discontinuity between university mathematics and school mathematics has proved to be extremely fruitful and can be seen, in both theoretical and practical respects, to constitute the core of mathematics teacher education. It remains an open question as to how much understanding the "elementary mathematics from an advanced standpoint" in the double discontinuity phenomenon can be achieved in teacher education, and to what extent it can solve the problems of discontinuity.

4 Transitions Between Teaching Institutions

4.1 Changes Between Educational Institutions at Different Levels of Determination

During their studies, students experience many transitions of educational institution: from pre-school to primary school, from primary to secondary school, sometimes from lower to higher secondary school or technical college, from higher secondary to university, from professional school or university to the workplace, etc. These transitions mean changes in many senses. First of all, and with some exceptions, a spatial one: another building, another school center, and sometimes another city and way of living. Transitions also include changes in the general functioning of the educational institution: number and types of teachers, number and types of students, time schedules, classroom equipment, etc. They can also affect pedagogic approach: relationships between teachers and students, kind of resources used, how learning goals and topics are specified and organized, etc. Finally, changes may occur in the kind of knowledge and know-how that is taught, as well as in the specific didactic activities related to this knowledge and know-how.

The study of the discontinuities found in the transition between educational institutions can be organized according to different levels of specificity, depending whether they affect the core learning and teaching content, which we will call the *didactic level*; whether they affect the organization of topics, which we will call the *pedagogic level*; or whether they affect the more global organization of learning and teaching activities in the educational institution, which we will call the *school level*, giving to school a broad sense of "place for instruction." In this respect, an interesting methodological tool proposed by Chevallard (2002) and also used in comparative studies (Artigue and Winsløw 2010) is the *scale of levels of didactic codetermination* (Fig. 2). The scale indicates that the way teaching and learning processes are organized determines, and is in turn determined by, conditions and constraints located at different levels of specificity. For instance, a more axiomatic or deductive organization of knowledge will favor and be reinforced by more "transmissive" pedagogies, which in turn will favor and be reinforced by traditional school organizations (one-hour lectures with one teacher and a large group of students). On the other hand, a change at the school level (for instance, to turn instruction into a more inquiry-based activity) needs changes not only at the pedagogical level (responsibilities assumed by the teacher and the students, for instance) but also at the more specific levels—how the different bodies of knowledge should be organized and presented.

Recognizing the various levels of the scale will help organize research findings on transitions found in the literature about mathematics education. We focus on the

Fig. 2 Scale of levels of didactic codetermination (from Artigue and Winsløw 2010, p. 52)

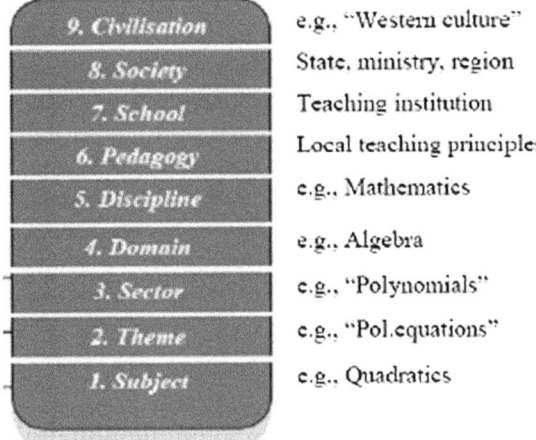

9. Civilisation	e.g., "Western culture"
8. Society	State, ministry, region
7. School	Teaching institution
6. Pedagogy	Local teaching principles
5. Discipline	e.g., Mathematics
4. Domain	e.g., Algebra
3. Sector	e.g., "Polynomials"
2. Theme	e.g., "Pol.equations"
1. Subject	e.g., Quadratics

primary-secondary and secondary-university transitions and which levels are questioned with respect to their desirable properties, which ones taken for granted, and how this depends on the kind of transition involved.

We will here discuss transitions to consider a change at the level of school and in relation to the teaching and learning of a specific discipline or field of knowledge (mathematics, in our case). It is important to take into account that the transition happens between institutions maintaining a certain ordered relationship, meaning that the first one is partially aiming at preparing the passage to the other: in many countries primary school prepares students to enter secondary school, secondary school to a technical college or university, and technical college and university to the workplace. A certain "grading" also exists with regards to the mathematics (or other field of knowledge) involved in the global educational process: many university teachers are researchers in mathematics or related fields of knowledge (engineering, sciences, etc.), closer to the production, development, and dissemination of mathematics than their colleagues at secondary level; these, in turn, have people who have usually taken more advanced mathematics studies than primary school teachers. It appears that these different positions can explain some asymmetry in the factors taken into account by researchers to explain the gaps found in the transitions.

We will now briefly consider some of the main research findings in the study of difficulties, discontinuities, or gaps found in the transition between, on the one hand, primary and secondary schools (P-S) and, on the other, between secondary school and university (S-U). We will try to locate the relevant phenomena at the different levels of the scale of didactic codetermination, in order to identify different patterns used in the treatment of P-S and S-U transitions. Even if most of the research in the literature proposes ways to bridge or smooth the discontinuities between schools, it is also important to notice that discontinuities are inherent to educational processes. Because their aim is to make students' knowledge and know-how evolve, the receiving schools need to highlight the changes made by exposing the differences with the feeding schools, a phenomenon that is especially visible at the frontier between institutions, where the transition takes place. Clark and Lovric (2008) propose to approach this issue in terms of "rites of passage," that is, the events and activities that assist the person undergoing it to achieve necessary changes in order to pass from one "community" (or school) to another. Even if transitions and the discontinuities they suppose can be approached as crucial (and not necessarily negative) crises—we do not have to forget that they can involve critical gains as well as losses—they are often the black spot where educational trajectories are interrupted, with special damage to the weakest social groups.

4.2 The Primary-Secondary Transition

When we look at investigations that approach the transition between primary to secondary compulsory education (for instance, Attard 2010) as well as pilot initiatives carried out in various countries or regions (see for instance Bryan and Treanor 2007 for the Scottish case), we find a common agreement about the influence, not always negative, of the following changes. At the school level, we pass from a main generalist teacher to various specialist teachers per class, with less fluent interaction between pupils and teachers, tighter time schedules, less relevant out-of-school activities, and a smaller intervention of parents in school events. At the level of pedagogy, hands-on materials and concrete activities are left behind, learning activities become more usually based in written activities and more transmission oriented, and less place is left for collective work and multidisciplinary activities.

When considering the level of the discipline, and with respect to the S-U transition, there are not many investigations focusing on the more specific levels, those related to a mathematical domain or theme. The main exception is research on early algebra (Carraher and Schliemann 2014). As a matter of fact, algebra has long been the "transition topic" par excellence, marking the frontier between elementary and secondary education. We can thus interpret research on early algebra as the first attempt to blur the frontier by introducing a properly secondary content at primary school. How this might affect (and be affected by) other mathematical domains, for instance measure and quantities, statistics, or geometry, is an open question. Geometry is the other domain that is approached in the P-S transition (see, for instance, Sdrolias and Triandafillidis 2008). Research on this area shows the impact of curriculum discontinuities on students' and teachers' practices.

In fact, many of the proposals to smooth the transition or to increase the engagement in mathematics of students entering secondary schools are mainly based on strengthening the relationships between primary and secondary teachers, sharing teaching activities carried out in both institutions, and, particularly, promoting more open activities during the first years of secondary school. Apart from early algebra, we do not find proposals to modify the curriculum (discipline level) or at least some areas or domains that are usually studied at both the end of primary school and at lower secondary. In summary, students' disengagement or adaptation difficulties are mainly attributed to pedagogical and school characteristics of secondary schools, beyond the discipline level. Therefore, the proposals to bridge the gap tend to concern this second level of instruction by questioning its pedagogical practices and trying to link them to those of primary school. Pressures for change always go in the direction of bringing lower secondary education closer to elementary education.

In many countries, compulsory education has been extended to the age of 15 or 16, so the passage from primary to secondary school is no longer elective. Generally, the expansion has been accomplished by combining a slight variation of the old non-compulsory education with basic education. Some curricular changes have also been made, but the general structure of content is not really modified. Therefore, the P-S transition remains a vestige of the time when only a minority of pupils pursued their studies while the large majority went directly to the workplace or, at most, to a vocational school. In some countries the primary level of studies is also called basic or elementary education since it involves the necessary preparation for any citizen to live in society. However, this idea has not necessarily been extended to compulsory secondary education, which remains in an uncomfortable position between education for all and preparation for the post-compulsory education of a few.

4.3 The Secondary-University Transition

The transition between secondary and university education has attracted the attention of many researchers during the last 20 years, as shown in Gueudet (2008) and Thomas et al. (2015). Again we will use the scale of levels of codetermination to present the main results and, more concretely, the phenomena observed as well as the "blind spots" that remain unquestioned:

At the school level, the passage from secondary school to university is generally marked by an increase in class size, having more than one teacher per subject, teachers that are also researchers with more time to prepare teaching but less to cooperate on pedagogical issues. Universities also provide richer equipment in written and technological resources that students are supposed to use more freely. At the pedagogical level, students find an increase in autonomy, a more transmission-based pedagogy requiring a more proactive use of resources (several books, articles, and notes in contrast to a unique textbook), while lecturers have more freedom in the organization of content and a greater variety in the types of assessments.

When we consider the level of the discipline, investigations characterize university mathematics as being more focused on the theoretical organization of mathematical content, the foundations of knowledge, and presenting proofs and theorems as tools to approach problems. In contrast, secondary mathematics stresses the production of results and the practical aspect of mathematical activities, assigning a more "decorative" role to axioms, definitions, and proofs. Bosch et al. (2004) talk about "incompleteness" of secondary school mathematics in the sense that it appears as a set of relatively isolated types of tasks with usually a single technique to solve them, the validation of which is more based on evidence than a formal construction. At the university, we can also speak of incompleteness in another sense: a lot of theoretical results are introduced without much connection to the problems and tasks that motivate them. The need for new knowledge is rarely associated with the content and practices of secondary school mathematics through, for instance, the consideration of mathematical phenomena that may require formal justification. Much research on the S-U transition approaches specific mathematical domains or themes. Calculus (functions, limits, and derivatives) is one of the most studied domains. Like arithmetic and algebra in the primary-secondary transition, the passage from calculus to analysis is paradigmatic of the entrance to "higher mathematics." Linear algebra and geometry are other domains that are critically involved in the transition, where the evidence based on geometrical facts in S needs to become formal proof in U.

Surprisingly, some of the phenomena pointed out as difficulties in the S-U transition appear in the same sense in the case of the P-S transition: more interaction between teacher and students in the lower level than in the new one, increase in the students' autonomy, passage from more active to more transmissive pedagogies, and stronger separation between disciplines (in P-S) or between domains within a discipline (in S-U). However, the treatment of these difficulties is clearly asymmetric: in the P-S transition it is the (higher) secondary school practice that requires change, while students' difficulties in the passage from S to U are also proposed to be overcome by changes in S (the lower level), as if the teaching and mathematical organizations at the university were unquestionable and thus untouchable. As an example, Hong et al. (2009) discuss secondary teachers' lack of knowledge of how calculus is taught at the tertiary level, without any inquiry or questioning about the lecturers' knowledge (and concern) about what is done at secondary level.

Research on the S-U transition has also focused on "bridging courses" organized in various universities to smooth the gap between upper-secondary school and university (Kayander and Lovric 2005; Biehler et al. 2011). Some of these studies (Serrano et al. 2007; Sierpinska et al. 2008) show how the bridging courses can actually increase the gap between institutions instead of facilitating the entrance to the new culture and its ways of working. This is the case for some courses proposing intensive work based on completing the required basic knowledge, thus reinforcing and rigidifying the old relationships to the old knowledge. In a sense, the bridging courses appear as a *coup de force* of the tertiary institution to clearly establish the entrance requirements of the new students, without any attempt to adapt its own practices to the newcomers and the feeding institutions. As indicated by Clark and Lovric (2008), they are not good devices for the "rite of passage" since they are rarely offered to all the students and are ineffective in giving them tools and confidence to approach the crisis better prepared.

4.4 Questions to Be Pursued

It is important to stress that, even if much research on transitions point to phenomena related to the general levels of codetermination (pedagogy, school, and society), approaching the specific levels is unavoidable since it is through the mathematical content that student-teacher relationships are built and teaching and learning activities concretized. In this respect, research on transitions needs to focus on broad domains of mathematics, covering more than one educational level. This enlarged unit of analysis facilitates raising questions that touch on the rationale of mathematical topics (what they are for), the reasons they should be taught, and even their delimitation (what they are). In this respect, the position of the researcher approaching transitions between teaching institutions appears to be a critical issue. What institutional point of view is assumed and which one is contested? Regarding mathematics (specific levels) we need to be cautious with the university perspective since it tends to appear as the most legitimate to secondary and primary mathematics teachers, thus reinforcing the propaedeutic function of the first levels of education (to facilitate the entrance to university mathematics) over their role in preparing students for citizenship.

5 Transitions Between In- and Out-of-School Mathematics

5.1 Introduction

For a very long time, mathematics educators and researchers in mathematics education have considered mathematics as universal and culture-free knowledge and have focused on the classroom as the primary—or even sole—setting in which mathematics learning takes place. Since the 1980s, numerous studies have documented that much mathematical knowledge is practiced, acquired, and transmitted outside school (Nunes 1992). This realization has brought new issues and questions about the transitions between these out-of-school mathematical practices and school mathematics into the research and practice of mathematics education.

Since 2000, this field of research on the existence of and transitions between various cultures of mathematics education has further evolved from "ethnomathematics and everyday cognition" (Nunes 1992) to "a broadening of the field, clarification and evolution of definitions, recognition of the complexity of the constructs and issues, and inclusion of social, critical, and political dimensions as well as those from cultural psychology, involving valorization, identity, and agency" (Presmeg 2007, p. 436).

This contribution provides a brief overview of some main themes, questions, findings, and recommendations relating to the multifold transitions between in- and out-of-school mathematics, with special attention to learners of early and elementary mathematics. We start by reviewing the state of the art in research on the transition from prior-to-school to school mathematics. Then we look at the research on the transitions from out-of-school to school mathematics (and vice versa).[1]

5.2   The Transition from Prior-to-School to School Mathematics

The transition from prior-to-school to school mathematics may be studied from a cognitive-developmental or a socio-cultural perspective.

Inspired by developments in the field of neuroscience (e.g., Butterworth 2015), the past two decades have witnessed the emergence of a very productive and influential line of cognitive research on children's early number sense, its development, and its relation to school mathematics. Cross-sectional and longitudinal studies have demonstrated that various core elements of children's early mathematical ability—such as their numerical magnitude understanding, their subitizing and counting skills, and their ability to transcode a number from one representation to another—are positively related to concurrent and future mathematics achievement (Torbeyns et al. 2015). More recent research is yielding increasing evidence for significant relations also with (1) young children's abilities related to mathematical relations, patterns, and structures and (2) their tendency to spontaneously attend to numerosities and to mathematical relations, patterns, and structures in their environment (Torbeyns et al. 2015). Furthermore, researchers have tried to enhance children's transition from informal to formal school mathematics by means of (computer-based) intervention programs aimed at enhancing one or more early numerical competencies—before or in the transition to formal mathematical instruction (Butterworth 2015; Torbeyns et al. 2015). While most of these interventions resulted in positive effects on the development of the trained early mathematical skills, evidence for positive transfer to mathematics learning more broadly is much less. The most convincing results have been obtained through long-lasting and broadly conceived intervention programs (e.g., Clements and Sarama 2014).

In a complementary line, researchers approach the transition from prior-to-school to school mathematics from a broader socio-cultural perspective wherein it is primarily conceived as a set of processes whereby individuals "cross borders" from one cultural or more specifically educational context or community to another. An overview of this perspective is provided by Perry et al. (2015), which points to the following issues.

---

[1]While the first kind of transition may be considered a non-reversible process, the second kind may, in contrast, be construed as an interaction—a "continuous transition" between two contexts (see Sect. 1).

First, prior to starting learning mathematics in school, children already engage in a wide range of early mathematical experiences and develop many sophisticated and powerful early mathematical ideas. While some of these experiences are explicitly linked to school mathematics, many others are less salient or more implicit and thus often go unrecognized as contributing to the mastery of mathematical skills by parents and teachers. Mathematical ideas that have already been developed may also not be revealed by traditional (collectively administered and worksheet-based) forms of assessment but require appropriate conversational interviewing techniques and meaningful tasks to elicit them.

Second, this research has provided ample evidence of the enormous variation in the frequency and the quality of exposure to early mathematical activities and experiences—both within and between cultures—which has shown to be predictively related to children's later mathematics achievement at school. For instance, using a large and representative sample of kindergartners from the United States, Galindo and Sheldon (2012) found that, on average, family involvement at school and parents' educational expectations were correlated with gains in mathematics achievement in kindergarten. Interestingly, this study also revealed that schools' efforts to communicate with and engage families predicted greater family involvement in school and, consequently, higher levels of student achievement in mathematics at the end of kindergarten. To give an example at the intercultural level, the Chinese cultural aspiration for academic success and the belief about the importance of an early start is generally considered an important explanatory factor for the excellent mathematical achievement scores of Chinese learners in international comparisons such as TIMSS.

Third, several qualitative studies have documented the serious transition problems that may arise when families, pre-school educators, and elementary school teachers are not working together to support the mathematics learning of children making the transition to school mathematics.

5.3   Transitions from Out-of-School to School Mathematics (and Vice Versa)

During the past decades many studies have looked at the transition from out-of-school to school mathematics (and vice versa). Three main themes of this literature are: (1) exploration of out-of-school mathematical practices and cultures (in comparison to mathematics learnt at school), (2) difficulties in the transition between out-of-school and school mathematics, and (3) attempts to facilitate these transitions.

As stated above, since the early works on ethnomathematics and everyday mathematics from the 1980s (Nunes 1992), researchers have continued to explore and describe forms of informal mathematics that are embedded in everyday cultural activities (Nunes 1992; Presmeg 2007). A first and very important perspective on these issues comes from research that has analyzed the forms and functions of mathematical activity in remote, indigenous, non-Western cultures where no systematic transmission in school prevails. Illustrative for this perspective is are Saxe's (2015) in-depth analyses of the Oksapmin numeration system based on body parts and of how certain changing socio-economic conditions (such as increased participation in economic exchanges involving currency) lead to important adaptations in that system. A second perspective involves analyses of informal mathematical practices that are embedded in specific out-of-school activities and contexts that may be contrasted with "school mathematics," as in the famous pioneering study of Brazilian street venders (see Nunes 1992). Continuing efforts have further contributed to our insight into the forms and social functions of mathematical activity demonstrated by people in varied out-of-school or, more specifically, work contexts such as carpet laying, interior design, retailing, restaurant management, dietetics, newspaper selling, nursing, banking, and architecture. These everyday math studies show that these forms and functions of mathematical activity are quite different from those of school mathematics, as they are "embodied in expressive forms and bodily modalities, distributed to other people and technologies, and are embedded in the language of the locals" (Reed 2013, p. 75).

While many of these analyses have focused on how people *perform* certain tasks, others have focused more on the mathematics *learning processes* in those out-of-school environments, showing that mathematics learning is certainly not limited to acquisition of the mathematical knowledge and skills passed down by mathematics teachers to individuals via school, but that it occurs as well during participation in cultural practices as people attempt to accomplish pragmatic goals. More specifically, these studies contributed to the view of mathematics learning in many of these out-of-school environments as "a centripetal movement of the apprentice from the periphery to the center of practice, under the guidance of those who are already masters in that practice" (Presmeg 2007, p. 444).

Many of these analyses of out-of-school mathematical practices and learning environments involve more or less explicit contrasts between doing and learning mathematics in these out-of-school contexts versus formal school contexts, glamorizing the former as more "authentic," "meaningful," "flexible," and/or "effective" than the latter. However, during the past decades, better understanding of the similarities and differences between these different contexts of mathematical (learning) practices has increasingly discouraged researchers and educators from making caricature-like characterizations of these different contexts of mathematical practices and, consequently, oversimplified recommendations for implementing features of these informal mathematical practices into mathematics classroom contexts.

Second, researchers have analyzed people's use of their mathematical knowledge and skills acquired in one context when functioning in the other. This analysis can be done in both directions.

As far as the transition from school to out-of-school mathematics is concerned, many researchers have documented—for elementary school up to university—(traditionally schooled) learners' difficulties in spontaneously and efficiently applying their formal mathematical knowledge and skills learned at school in relevant out-of-school contexts (e.g., C&TG at Vanderbilt 1997).

Rather than looking at school mathematics' lack of transfer to real life and work situations, various other lines of research have inversely focused on how real-world knowledge does—or rather does not—permeate into school mathematics. One such line of research has documented learners' reluctance or inability to make productive use of their experiential knowledge and realistic considerations about the real world situation described in the problem when solving word problems in the mathematics classroom. As shown by this research, the practice and culture of the word problem-solving lessons at school develop the belief in learners that there is a huge gap between solving a word problem at school and solving a comparable problem in the real world outside school, making it counterproductive to rely on one's experiential knowledge about the problem situation (Verschaffel et al. 2009). Whereas the above research is concerned with learners' reluctance to allow relevant real-world knowledge and sense-making into their word problem-solving endeavors at school, other research has analyzed the restrictions of people's everyday mathematical knowledge and skills vis-à-vis academic mathematical tasks (such as mathematical word problems). For instance, Nunes' (1992) analyzed unschooled adults' ability to solve *indirect* word problems (i.e., problems that needed the application of the mathematical inverse principle, e.g., $a + b = c$ so $c - b = a$) and showed how difficult it was for them to solve such problems based on their everyday mathematical knowledge and skills. These difficulties were explained in terms of the highly contextualized nature of the mathematical knowledge and skills acquired by these adults in informal learning environments, whereas mathematical competencies acquired at school are assumed to be of a more general and abstract nature. This is not to say that the situated mathematical knowledge and skills acquired out of school have no generalizability at all; they do, but it typically remains restricted to their specific context of application (e.g., they do not transfer beyond the carpenter's working context).

Third, researchers have looked—at different levels of specificity and again in both directions—for ways to bridge the gap between in- and out-of-school mathematics.

At a general level, many innovative approaches to the teaching of elementary and secondary school mathematics, have—as one of their major design principles— striven to bring out-of-school problems, experiences, and practices into the arena of school mathematics. Probably one of the most well-known examples is the Realistic Mathematics Education approach, which starts from the basic idea that, rather than having learners first acquire the formal system of mathematics with the applications to come afterwards, mathematical knowledge should be gathered and developed starting from the exploration and study of phenomena in the real world (Treffers 1987). In the past decades, many other scholars have developed and implemented similar educational approaches wherein mathematical concepts and skills are developed starting from a real or imaginable out-of-school situation and linking it in a number of steps with formal school mathematics (Presmeg 2007).

At a more specific level, particularly in the domain of mathematical modeling, many researchers have documented various kinds of efforts—from the first years of elementary school to the university level—aimed at making mathematical modelling less scholastic and more realistic. These studies vary from basic attempts to increase the authenticity of the formulation and presentation of (standard) word problems to more drastic efforts to replace these word problems with challenging real-world mathematical modeling problems presented and embedded in rich and multi-componential technological environments (e.g., C&TG at Vanderbilt 1997). While several of these studies have reported successes in terms of learner outcomes, they also point to many difficulties—mathematical, pedagogical and managerial—that occur when teachers try to implement these novel modelling tasks and their accompanying pedagogies in the mathematics classroom (Verschaffel et al. 2009).

Finally, as in the above-mentioned studies with young children in the transition from prior-to-school to school mathematics, other researchers have tried to strengthen—from a socio-cultural rather than a cognitive-psychological perspective—the transitional links between learners' home and school culture. They do so by setting up productive forms of home-school collaboration with a view to help parents of learners—typically learners with a minority or immigration background—to support their children in the transition to learning mathematics in a school culture that is new for them. Prominent examples are the Home-School Knowledge Exchange Project by Hughes and Pollard (2006) in the United Kingdom and the BRIDGE: Linking Home and School project by Civil and Andrade (2002) in the United States.

In general, the transition between in- and out-of-school mathematics continues to be an issue of great theoretical and practical importance, even after learners have made the important cognitive and cultural step from preschool to school mathematics. This transition issue is also approached from different and complementary theoretical angles—cognitive-psychological as well as socio-cultural—in both directions, and at various levels of specificity. The extremely dichotomous descriptions of the features and merits of in- and out-of-school (learning) practices have been replaced by more nuanced and complex descriptions and analyses of these different kinds of mathematical practices (and the various kinds of intermediate mathematical practices) and of the various types of transitions (in both directions) between them.

6 Summary and Looking Ahead

The five chapters of this survey have only sketched some essential aspects of the very rich body of international research in mathematics education addressing transitions. The reader interested will find below a complete references list and some selected texts for further reading.

How should we proceed to further develop the research reviewed in our synthesis?

- *The need for combining approaches*

Our aim was to order the discussion of the body of relevant literature by separating the historical, cognitive, epistemological, and socio-cultural approaches. At the end of this survey, it appears clearly that most of the research engages several of these approaches. Our survey has evidenced the complementarity of the different views on transition, particularly the cognitive and socio-cultural views. Studying the complex evolutions taking place during educational transitions probably requires drawing on several theoretical perspectives, offering different lenses and also certainly diverse associated methods. Analyses of students' discourse, comparison of mathematical texts provided by two different institutions, specific interviews with young children, didactical engineering, etc., have been used and sometimes combined by the authors of the works we cited. One of the challenges for future research on transition is to build methods that permit not only the analysis of the initial and the final state but also allow grasping the process of change itself in all its complexity in order to provide students and teachers' resources that foster this process.

- *From gaps and obstacles to commonalities and opportunities*

The research discussed in the state-of-the-art synthesis above has mainly addressed discontinuities and difficulties attached to transitions, such as epistemological obstacles hindering conceptual change, gaps between out-of-school and in-school mathematics, and chasms between institutions. In more recent works we see some interesting evolution that can suggest directions for further research. These works propose more balanced analyses where transition is not depicted as a route paved with obstacles only, but as a complex process where difficulties are also associated with opportunities. Concerning conceptual change, we propose that the transition from naïve to expert knowledge can correspond to a variety of paths, which must be searched out and explored. For teachers, subject matter knowledge and pedagogical content knowledge are not independent. About institutional transitions, we observed that not only differences but also common features appear across transitions in institutions and that these common features are likely to be very informative about school mathematics in particular countries. The search for common elements, or a continuous path from out-of-school to school mathematics, is likely made more complex by the variety of possible contexts: family mathematics, which changes from one family to another; mathematics for nursing; mathematics for playing music, etc., are all different. Nevertheless communication between the actors in the different contexts can help to identify elements from which coherent learning paths can be built.

- *Research results and interventions on transitions*

Amongst the research results about transition, analyses of the situation in ordinary classes can usefully inform teachers and other educational agents at all school levels—including families. A first form of educational intervention drawing on research can be to address this important topic in teacher education and in journals for non-specialists.

Research on transitions has also proposed teaching devices: specific mathematical situations likely to foster conceptual change, bridging courses between different institutions, teacher training programs where university mathematics is used to build coherent teaching for secondary school, and courses about modeling where the "real world" and mathematics are connected.

For transitions between different contexts, developing communities where teachers (or other actors) can communicate and work together to elaborate a common vision of the teaching of mathematics (Barton et al. 2010) is a promising direction. In fact, for all kinds of transitions, the complexity of the phenomena taking place suggests that designing possible interventions requires a collective effort and collaborations among the members of all the sub-communities of the ICME.

References

Artigue, M., & Winsløw, C. (2010). International comparative studies on Mathematics education: A viewpoint from the Anthropological theory of didactics. *Recherches en Didactique des Mathématiques,31*(1), 47–82.

Arzarello, F., & Sabena, C. (2011). Meta-cognitive unity in indirect proofs. In M. Pytlak, T. Rowland, & E. Swoboda (Eds.), *Proceedings of the VIIth Conference of the European Society for Research in Mathematics* (pp. 99–109). Poland: Rzeszow.

Attard, C. (2010). Students' experiences of mathematics during the transition from primary to secondary school. In L. Sparrow, B. Kissane, & C. Hurst (Eds.), *Shaping the future of mathematics education* (pp. 53–60). Fremantle: MERGA.

Bachelard, G. (1938). *La formation de l'esprit scientifique*. Paris, France: Librairie Philosophique J. Vrin.

Ball, D. L., Hill, H. C., & Bass, H. (2005). Knowing mathematics for teaching: Who knows mathematics well enough to teach third grade, and how can we decide? *American Educator, 29* (1), pp. 14–17, 20–22, 43–46.

Ball, D. L., & Forzani, F. (2009). The work of teaching and the challenge for teacher education. *Journal of Teacher Education,60*(5), 497–511.

Barton, B., Clark, M., & Sherin, L. (2010). Collective dreaming: a school-university interface. *New Zealand Journal of Mathematics,40*, 15–31.

Biehler, R., Fischer, P. R., Hochmuth, R., & Wassong, Th. (2011). Designing and evaluating blended learning bridging courses in mathematics. In *Proceedings of the Seventh Congress of the European Society for Research in Mathematics Education* (pp. 1971–1980).

Biza, I., Jaworski, B., & Hemmi, K. (2014). Communities in university mathematics. *Research in Mathematics Education,16*(2), 161–176.

Blömeke, S., Suhl, U., & Kaiser, G. (2011). Teacher education effectiveness: Quality and equity of future primary teachers' mathematics and mathematics pedagogical content knowledge. *Journal of Teacher Education,62*(2), 154–171.

Blömeke, S., Hsieh, F.-J., Kaiser, G., & Schmidt, W. H. (Eds.). (2014). *International Perspectives on Teacher Knowledge, Beliefs and Opportunities to Learn*. Dordrecht: Springer.

Borko, H. (2004). Professional development and teacher learning: Mapping the terrain. *Educational Research,33*(8), 3–15.

Bosch, M., Fonseca, C., & Gascón, J. (2004). Incompletitud de las organizaciones matemáticas locales en las instituciones escolares. *Recherches en Didactique des Mathématiques,24*(2–3), 205–250.

Boston, M. D. (2012). Connecting changes in secondary mathematics teachers' knowledge to their experiences in a professional development workshop. *Journal of Mathematics Teacher Education,16*, 7–31.

Bryan, R., & Treanor, M. (2007). *Evaluation of pilots to improve primary to secondary school transitions*. Glasgow, United Kingdom: Scottish Executive Social Research.

Buchholtz, N., Leung, F. K., Ding, L., Kaiser, G., Park, K., & Schwarz, B. (2013). Future mathematics teachers' professional knowledge of elementary mathematics from an advanced standpoint. *ZDM, The International Journal on Mathematics Education,45*(1), 107–120.

Butterworth, B. (2015). Low numeracy: From brain to education. In X. Sun, B. Kaur, & J. Novotná (Eds.), *The twenty-third ICMI study: Primary mathematics study on whole numbers* (pp. 21–33). Macao, China: University of Macau.

Carey, S. (1985). *Conceptual change in childhood*. Cambridge, MA: MIT Press/Bradford Books.

Carey, S. (2009). *The origins of concepts*. Oxford, UK: Oxford University Press.

Carraher, D., & Schliemann A. D. (2014). Early Algebra Teaching and Learning. In S. Lerman (Ed.), *Encyclopedia of Mathematics Education*, (pp. 193–196). Springer.

Chassapis, D. (1998). The mediation of tools in the development of formal mathematical concepts: The compass and the circle as an example. *Educational Studies in Mathematics,37*(3), 275–293.

Chevallard, Y. (2002). Organiser l'étude 3. Écologie & régulation. In J. L. Dorier, M. Artaud, M. Artigue, R. Berthelot, & R. Floris (Eds.), *Actes de la 11e école de didactique des mathématiques* (pp. 41–56). Grenoble: La Pensée Sauvage.

Chevallard, Y. (2006). Steps towards a new epistemology in mathematics education. In M. Bosch (Ed.), *Proceedings of the IVth Conference of the European Society for Research in Mathematics Education* (pp. 22–30). Spain: Barcelona.

Chi, M. T. H. (1992). Conceptual change across ontological categories: Examples from learning and discovery in science. In F. Giere (Ed.), *Cognitive models of science: Minnesota studies in the philosophy of science* (pp. 129–160). Minneapolis: University of Minnesota Press.

Chi, M. T. H. (2013). Two kinds and four sub-types of misconceived knowledge, ways to change it and the learning outcomes. In S. Vosniadou (Ed.), *International handbook of research on conceptual change* (2nd ed., pp. 49–70). New York, NY: Routledge.

Civil, M., & Andrade, R. (2002). Transitions between home and school mathematics: rays of hope amidst the passing clouds. In G. de Abreu, A. J. Bishop, N. C. Presmeg (Eds.), *Transitions between contexts of mathematical practices* (pp. 149–170). Dordrecht: Kluwer.

Clark, M., & Lovric, M. (2008). Suggestion for a Theoretical Model for Secondary-Tertiary Transition in Mathematics. *Mathematics Education Research Journal,20*(2), 25–37.

Clements, D. H., & Sarama, J. (2014). *Learning and teaching early math: The learning trajectories approach* (2nd ed.). New York, NY: Routledge.

Cognition and Technology Group at Vanderbilt. (1997). *The Jasper project: Lessons in curriculum, instruction, assessment, and professional development*. Mawhaw, NJ: Erlbaum.

Conference Board on the Mathematical Sciences. (2012). *The Mathematics Education of Teachers II*. Providence, Rhode Island: The American Mathematical Society.

Crafter, S., & de Abreu, G. (2011). Teachers discussions about parental use of implicit and explicit mathematics in the home. In M. Pytlak, T. Rowland, & E. Swoboda (Eds.), *Proceedings of the VIIth Conference of the European Society for Research in Mathematics* (pp. 1419–1428). Poland: Rzeszow.

Crafter, S., & Maunder, R. (2012). Understanding transitions using a sociocultural framework. *Educational and Child Psychology,29*(1), 10–18.

diSessa, A. A. (2013). A bird's eye view of "pieces" vs. "coherence" controversy. In S. Vosniadou (Ed.), *Handbook of conceptual change research* (2nd ed.) (pp. 31–48). New York, NY: Routledge.

diSessa, A. A. (2014). The construction of causal schemes: Learning mechanisms at the knowledge level. *Cognitive Science,38*(5), 795–850.

Dorier, J.-L. (Ed.). (2000). *On the Teaching of Linear Algebra*. Dordrecht: Kluwer Academic Publishers.

Dubinsky, E. (1991). Reflective abstraction in advanced mathematical thinking. In D. Tall (Ed.), *advanced mathematical thinking* (pp. 95–126). Dordrecht: Kluwer.

Galindo, C., & Sheldon, S. B. (2012). School and home connections and children's kindergarten achievement gains: The mediating role of family involvement. *Early Childhood Research Quarterly,27*, 90–103.

Garuti, R., Boero, P., Lemut, E., & Mariotti, M. A. (1996). Challenging the traditional school approach to theorems: A hypothesis about the cognitive unity of theorems. *Proceedings of PME 20* Vol. 2 (pp. 113–120). Valencia, Spain.

Goodson-Espy, T. (1998). The roles of reification and reflective abstraction in the development of abstract thought: Transition from arithmetic to algebra. *Educational studies in mathematics,36*, 219–245.

Gueudet, G. (2008). Investigating the secondary-tertiary transition. *Educational Studies in Mathematics,67*(3), 237–254.

Herbst, P., & Kosko, K. (2014). Mathematical knowledge for teaching and its specificity to high school geometry instruction. In J.-J. Lo, K. R. Leatham, & L. R. Van Zoest (Eds.), *Research Trends in Mathematics Teacher Education* (pp. 23–45). Springer.

Hill, H., Ball, D. L., & Schilling, S. (2008). Unpacking "pedagogical content knowledge": Conceptualizing and measuring teachers' topic-specific knowledge of students. *Journal for Research in Mathematics Education,39*(4), 372–400.

Hill, H. C., Schilling, S. G., & Bal, D. L. (2004). Developing measure of teachers' mathematics knowledge for teaching. *The Elementary School Journal,105*(1), 11–30.

Hodgson, B. (2015). Whither the Mathematics/Didactics Interconnection? Evolution and Challenges of a Kaleidoscopic Relationship as Seen from an ICMI Perspective. In S. J. Cho (Ed.), *Proceeding of the 12th International Congress on Mathematical Education*. New York, NY: Springer.

Hong, Y. Y., et al. (2009). A comparison of teacher and lecturer perspectives on the transition from secondary to tertiary mathematics education. *International Journal of Mathematical Education in Science and Technology,40*(7), 877–889.

Hughes, M., & Pollard, A. (2006). Home–school knowledge exchange in context. *Educational Review,58*, 385–395.

Kaiser, G., & Buchholtz, N. (2014). Overcoming the Gap between University and School Mathematics. In S. Rezat, M. Hattermann, & A. Peter-Koop (Eds.), *Transformation—A Fundamental Idea of Mathematics Education* (pp. 85–105). New York: Springer.

Kayander, A., & Lovric, M. (2005). Transition from secondary to tertiary mathematics: McMaster University experience. *International Journal of Mathematical Education in Science and Technology,36*(2–3), 149–160.

Kendal, M., & Stacey, K. (2002). Teachers in transition: Moving towards CAS-supported classrooms. *ZDM, The International Journal on Mathematics Education,34*(5), 196–203.

Klein, F. (1908/1939). *Elementary Mathematics from an Advanced Standpoint. Part I: Arithmetic, Algebra, Analysis. Part II: Geometry.* (E. R. Hedrick & C. A. Noble, Trans.). New York: Dover Publications.

Klein, F. (1933). *Elementarmathematik vom höheren Standpunkte aus I: Arithmetik-Algebra-Analysis*. Berlin: Julius Springer.

Krauss, S., Blum, W., Brunner, M., Neubrand, M., Baumert, J., Kunter, M., et al. (2013). Mathematics teachers' domain-specific professional knowledge: Conceptualization and test construction in COACTIV. In M. Kunter, J. Baumert, W. Blum, U. Klusmann, S. Krauss, & M. Neubrand (Eds.), *Cognitive activation in the mathematics classroom and professional competence of teachers* (Vol. 8, pp. 147–174). New York Heidelberg Dordrecht London: Springer.

Kuhn, T. S. (1970). *The structure of scientific revolutions* (2nd ed.). Chicago, IL: University of Chicago Press.

Kwon, O., Kim, A., & Cho, H. (2012). An analysis of mathematics teacher education curriculum and teaching methods in Korea. *Journal of Korean Society of Mathematical Education (Mathematical Education), 41*(3), 275–294.

Lai, Y., & Weber, K. (2013). Factors mathematicians profess to consider when presenting pedagogical proofs. *Educational Studies in Mathematics,85*, 93–108.

Lewis, J. M., & Blunk, M. L. (2012). Reading between the lines: Teaching linear algebra. *Journal of Curriculum Studies,44*, 515–536.

Linn, M., Eylon, B.-S., & Davis, E. (2004). The knowledge integration perspective on learning. In M. C. Linn, E. Davis, & P. Bell (Eds.), *Internet environments in science education* (pp. 29–46). Mahwah, NJ: Lawrence Erlbaum Associates.

Loch, C., Lindmeier, A., & Heinze, A. (2015). The missing link? School-related content knowledge of pre-service mathematics teachers. In K. Beswick, T. Muir, & J. Wells (Eds.), *Proceeding of the 39th Conference of the International Croup for the Psychology of Mathematics Education* (Vol. 3, pp. 209–216). Hobart (Tasmania): PME.

Mason, J., & Spence, M. (1999). Beyond mere knowledge of mathematics: The importance of knowing-to act in the moment. *Educational Studies in Mathematics,38*, 135–161.

Minstrell, J. (1989). Teaching science for understanding. In L. B. Resnick & L. E. Klopfer (Eds.), *Toward the thinking curriculum: Current cognitive research* (pp. 129–149). Alexandria, VA: Association for Supervision and Curriculum Development.

Nesher, P., & Peled, I. (1986). Shifts in reasoning. *Educational Studies in Mathematics,17*(1), 67–79.

Nunes, T. (1992). Ethnomathematics and everyday cognition. In D. A. Grouws (Ed.), *Handbook of research on mathematics teaching and learning* (pp. 557–574). New York: Macmillan.

Parnafes, O., & diSessa, A. A. (2013). Microgenetic learning analysis: A methodology for studying knowledge in transition. *Human Development,56*(5), 5–37.

Perry, B., MacDonald, A., & Gervasoni, A. (Eds.). (2015). *Mathematics and transition to school. International perspectives*. Singapore: Springer.

Pratt, D., & Noss, R. (2002). The microevolution of mathematical knowledge: The case of randomness. *Journal of the Learning Sciences,11*(4), 455–488.

Presmeg, N. (2007). The role of culture in teaching and learning mathematics. In F. K. Lester (Ed.), *Second handbook of research on mathematics teaching and learning* (pp. 435–458). Greenwich, CT: Information Age Publishing.

Reed, S. (2015). What counts too much and too little as mathematics. In B. Bevan P. Bell, R. Stevens, & A. Razfar (Eds.), *LOST opportunities: Learning in our-of-school time*. London: Springer.

Ríordáin, M. N., & O'Donoghue, J. (2011). Tackling the transition—the English mathematics register and students learning through the medium of Irish. *Mathematics Education Research Journal,23*(1), 43–65.

Robert, A. (1998). Outils d'analyse des contenus mathématiques à enseigner au lycée et à l'université (Tools for the analysis of the mathematical content taught at high school and university). *Recherches en Didactique des Mathématiques,18*(2), 139–190.

Sadovsky, P., & Sessa, C. (2005). The adidactic interaction with the procedures or peers in the transition from arithmetic to algebra: A milieu for the emergence of new questions. *Educational studies in mathematics,59*, 85–112.

Saxe, G. B. (2015). Studying culture-cognition relations in collective practices of daily life: a research framework. *Infancia y Aprendizaje,38*, 1–36.

Serrano, L., Bosch, M., Gascón, J. (2007). Diseño de organizaciones didácticas para la articulación del bachillerato con el primer ciclo universitario. In L. Ruiz, A. Estepa, & F. J. García (Eds.), *Sociedad, escuela y matemáticas. Aportaciones de la Teoría Antropológica de lo Didáctico* (pp. 757–766). Jaén: Publicaciones de la Universidad de Jaén.

Schubring, G. (2011). Conceptions for relating the evolution of mathematical concepts to mathematics learning—epistemology, history, and semiotics interacting. *Educational Studies in Mathematics,77*(1), 79–104.

Sdrolias, K., & Triandafillidis, A. (2008). The transition to secondary school geometry: can there be a "chain of school mathematics". *Educational studies in mathematics,67*, 159–169.

Sfard, A. (1991). On the dual nature of mathematical conceptions: Reflections on processes and objects as different sides of the same coin. *Educational Studies in Mathematics,22*(1), 1–36.

Sfard, A. (2007). When the rules of discourse change, but nobody tells you: Making sense of mathematics learning from a commognitive standpoint. *The Journal of the Learning Sciences,16*(4), 565–613.

Shulman, L. S. (1986). Those who understand: Knowledge growth in teaching. *Educational Researcher,15*(2), 4–14.

Sierpińska, A. (1987). Humanities students and epistemological obstacles related to limits. *Educational Studies in Mathematics,18*(4), 371–397.

Sierpińska, A. (1990). Some remarks on understanding in mathematics. *For the Learning of Mathematics,10*(3), 24–36.

Sierpińska, A., Bobos, G., & Knipping, C. (2008). Sources of students' frustration in preuniversity level, prerequisite mathematics courses. *Instructional Science,36*, 289–320.

Skemp, R. R. (1962). The need for a schematic learning theory. *British Journal of Educational Psychology,32*(2), 133–142.

Slavit, D. (1999). The role of operation sense in transitions from arithmetics to algebraic thought. *Educational studies in mathematics,37*, 251–274.

Tall, D. (2002). Continuities and discontinuities in long-term learning schemas. In D. Tall & M. Thomas (Eds.), *Intelligence, learning and understanding: A tribute to Richard Skemp* (pp. 151–177). Flaxton QLD, Australia: Post Pressed.

Tall, D., Thomas, M., Davis, G., Gray, E., & Simpson, A. (1999). What Is the Object of the Encapsulation of a Process? *The Journal of Mathematical Behavior,18*(2), 223–241.

Thomas, M. O. J., De Freitas Druck, I., Huillet, D., Ju, M.-K., Nardi, E., Rasmussen, C., & Xie, J. (2015). Key mathematical concepts in the transition from secondary to university. In S. J. Cho (Ed.), *The Proceedings of the 12th International Congress on Mathematical Education* (pp. 265–284). New York: Springer.

Thompson, P. W. (2013). In the absence of meaning. In K. Leatham (Ed.), *Vital directions for research in mathematics education* (pp. 57–93). New York: Springer.

Thompson, P. W. (2015). Researching mathematical meanings for teaching. In L. D. English & D. Kirshner (Eds.), *Third handbook of international research in mathematics education* (pp. 968–1002). New York: Taylor & Francis.

Torbeyns, J., Gilmore, C., & Verschaffel, L. (Eds.) (2015). The acquisition of preschool mathematical abilities: Theoretical, methodological and educational considerations. An introduction. *Mathematical Thinking and Learning, 17*, 99–115.

Toulmin, S. (1972). *Human understanding* (Vol. 1). Oxford, UK: Clarendon Press.

Treffers, A. (1987). *Three dimensions. A model of goal and theory description in mathematics education*. Dordrecht: Kluwer.

Verschaffel, L., Greer, B., Van Dooren, W., & Mukhopadhyay, S. (Eds.). (2009). *Words and worlds: Modelling verbal descriptions of situations*. Rotterdam: Sense.

Vosniadou, S. (2013). Conceptual change in learning and instruction: The framework theory approach. In S. Vosniadou (Ed.), *International handbook of conceptual change* (2nd ed., pp. 11–30). New York, NY: Routledge.

Wagner, J. F. (2006). Transfer in pieces. *Cognition and Instruction,24*(1), 1–71.

Wenger, E. (1998). *Communities of practice: Learning, meaning and identity.* Cambridge, UK: Cambridge University Press.

Winsløw, C. (2014). The transition from university to high school and the case of exponential functions. In B. Ubuz, C. Haser, M. A. Mariotti, *Proceedings of the Eighth Congress of the European Mathematical Society for Research in Mathematics Education* (pp. 2476–2485). Antalya, Turkey.

Winsløw, C., & Grønbæck, N. (2014). Klein's double discontinuity revisited: contemporary challenges for universities preparing teachers to teach calculus. *Recherches en Didactique des Mathématiques,34*(1), 59–86.

Wiser, M. (1995). Use of history of science to understand and remedy students' misconceptions about heat and temperature. In D. Perkins, J. Schwartz, M. West, & M. Wiske (Eds.), *Software goes to school: Teaching for understanding with new technologies* (pp. 23–38). New York; Oxford, UK: Oxford University Press.

Wiser, M., & Carey, S. (1983). When heat and temperature were one. In D. Gentner & A. L. Stevens (Eds.), *Mental models* (pp. 267–297). Hillsdale, NJ: Lawrence Erlbaum Associates.

Yerushalmy, M. (2005). Challenging known transitions: Learning and teaching algebra with technology. *For the Learning of Mathematics,25*(3), 37–42.

Further Reading

We present here central references, with a short informative comment for each.

Attard, C. (2010). Students' experiences of mathematics during the transition from primary to secondary school. In L. Sparrow, B. Kissane, & C. Hurst (Eds.), *Shaping the future of mathematics education* (pp. 53–60). Fremantle: MERGA.

Compared to the secondary-tertiary transition, the primary-secondary one has received less attention in the specific case of mathematics. Here is an interesting study which, however, remains at a general level, without considering specific mathematical learning contents.

de Abreu, G., Bishop, A. J., & Presmeg, N. C. (Eds.). (2002). *Transitions between contexts of mathematical practices.* Dordrecht: Kluwer.

This edited book offers both empirical studies and theoretical reflections on mathematics learners in transition and on their practices in different contexts, mainly from a socio-cultural perspective.

diSessa, A. A. (2014). A history of conceptual change research: Threads and fault lines. In K. Sawyer (Ed.), *Cambridge handbook of the learning sciences* (2nd ed., pp. 88–108). Cambridge, UK: Cambridge University Press.

This chapter attempts a balanced overview of the history of conceptual change research, although it is written by an active participant in ongoing debates.

Gueudet, G. (2008). Investigating the secondary-tertiary transition. *Educational Studies in Mathematics,67*(3), 237–254.

This article provides an overview on the interface between secondary school and tertiary institution.

Klein, F. (1908/1939). *Elementary Mathematics from an Advanced Standpoint. Part I: Arithmetic, Algebra, Analysis. Part II: Geometry.* (E. R. Hedrick & C. A. Noble, Trans.). New York: Dover Publications.

This volume contains Klein's idea on double discontinuity mentioned in Sect. 1.3.

Perry, B., MacDonald, A., & Gervasoni, A. (Eds.). (2015). *Mathematics and transition to school. International perspectives.* Singapore: Springer.

This edited book brings together an international collection of work built around two important components of any young child's life: learning mathematics and starting elementary school.

Thomas, M. O. J., De Freitas Druck, I., Huillet, D., Ju, M.-K., Nardi, E., Rasmussen, C., & Xie, J. (2015). Key mathematical concepts in the transition from secondary to university. In S.

J. Cho (Ed.), *The Proceedings of the 12th International Congress on Mathematical Education* (pp. 265–284). New York: Springer.
An interesting survey about research on secondary-tertiary transition, with a focus on different perspectives and mathematical areas.
Vosniadou, S. (Ed.). (2013). *International handbook of research on conceptual change* (2nd ed.). New York, NY: Routledge.
This volume contains up-to-date descriptions of conceptual change research, including chapters by many of the protagonists mentioned in Chapter 2.

www.ingramcontent.com/pod-product-compliance
Ingram Content Group UK Ltd.
Pitfield, Milton Keynes, MK11 3LW, UK
UKHW020217231225
466357UK00011B/187